discovermore
Exploring Physical Science

Measurement

Marie Harts

Britannica
Educational Publishing
IN ASSOCIATION WITH
THE ROSEN PUBLISHING GROUP

Published in 2024 by The Rosen Publishing Group, Inc.
2544 Clinton Street, Buffalo, NY 14224

Editor: Shannon Harts
Book Design: Rachel Rising

Photo Credits: Cover yanishevska/Shutterstock.com; (series background) Dai Yim/Shutterstock.com; p.4 yurakrasil/Shutterstock.com; p. 5 anut21ng Stock/Shutterstock.com; p. 6 Sergey Nivens/Shutterstock.com; p. 7 Blue Green Eyes/Shutterstock.com; p. 8 Hitra/Shutterstock.com; p. 9 charnsitr/Shutterstock.com; p. 9 Gil C/Shutterstock.com; p. 9 Mania Mee/Shutterstock.com; p. 10 Sternfahrer/Shutterstock.com; p. 11 Shutter.B/Shutterstock.com; p. 13 grandbrothers/Shutterstock.com; p. 13 Joa Souza/Shutterstock.com; p. 14 DC Studio/Shutterstock.com; p. 15 Frogastro/Shutterstock.com; p. 17 fizkes/Shutterstock.com; p. 17 Zyn Chakrapong/Shutterstock.com; p. 18 PeopleImages.com - Yuri A/Shutterstock.com; p. 19 mpohodzhay/Shutterstock.com; p. 20 Graham De'ath/Shutterstock.com; p. 21 sockagphoto/Shutterstock.coml; p. 22 Africa Studio/Shutterstock.com; p. 23 r.classen/Shutterstock.com; p. 24 Hanasaki/Shutterstock.com; p. 25 BongkarnGraphic/Shutterstock.com; p. 26 sergioluno/Shutterstock.com; p. 27 Andrei Armiagov/Shutterstock.com; p. 28 Prostock-studio/Shutterstock.com; p. 29 Gorodenkoff/Shutterstock.com.

Cataloging-in-Publication Data

Names: Harts, Marie.
Title: Measurement / Marie Harts.
Description: New York : Britannica Educational Publishing, in Association with Rosen Educational Services. 2024. | Series: Discover more: exploring physical science | Includes glossary and index.
Identifiers: ISBN 9781641900607 (library bound) | ISBN 9781641900591 (pbk) | ISBN 9781641900614 (ebook)
Subjects: LCSH: Measurement--Juvenile literature.
Classification: LCC QA465.H37 2024 | DDC 530.8--dc23

Manufactured in the United States of America

CPSIA Compliance Information: Batch #CWBRIT24. For further information contact Rosen Publishing at 1-800-237-9932.

Contents

What Is a Measurement?

If you're trying to figure out the size, amount, or degree of something, you're taking a measurement. Many things can be measured, including length, area, volume, angles, weight, temperature, and time.

Measuring is important for many jobs, including for seamsters and tailors who take measurements to mend clothing.

Construction workers must know how to take accurate measurements because it can affect the quality and safety of what they're building.

Different measurements require different tools. For example, to measure the length of a piece of wood, you could use a ruler. But to measure temperature, you would use a thermometer. Different units are used too. Length can be measured in meters (in the metric system) or feet (in the imperial system). There are also smaller and larger units. For example, a centimeter is a smaller unit than the meter used in the metric system, and an inch is a smaller unit than the foot used in the imperial system.

Consider This

What tool might you use to measure your height? What about your body temperature when you're sick? Measurements can shape our understanding of ourselves daily!

Understanding Standards

Society depends on standard measurements, which allow one person to tell another person a measurement's value without confusion. A standard is an amount upon which everyone agrees. Units are the names for different standards. For example, one unit for weight is the gram. Everyone agrees that 1 gram is a certain amount of weight. Two grams equal twice that amount, and so on.

In the 1950s, scientists decided to measure time based on the vibrations of atoms, which are the smallest bits of matter, because these vibrations will never change.

A solar day is a measurement of the amount of time between noon one day on Earth to the next day. Noon is defined as the time the sun is at its highest point in the sky.

Some standards are based on fundamental laws of nature. For instance, the meter is defined as the distance traveled by light in a vacuum during a period of 1/299,792,458th of a second. A second used to be based on 1/86,400th of a solar day. However, as the average solar day is getting longer, a more stable way to define the second was needed. Today, a second is based on the time it takes for an atom, often cesium, to vibrate 9,192,631,770 times.

Consider This

Why do you think a non-varying, or constant, physical event is important for defining measurement standards?

Investigating the Imperial System

Before the French Revolution in 1789, just about every time people traveled to a new country, they had to learn a new standard measurement system. Today, people usually use one of two systems: the imperial system or the metric system.

In the imperial system, the inch is a unit of length. There are 12 inches in 1 foot and 5,280 feet in 1 mile. The cup is a unit of volume. There are 2 cups in 1 pint, 2 pints in 1 quart, and 4 quarts in 1 gallon. The ounce is the unit of weight. There are 16 ounces in a pound and 2,000 pounds in a ton.

In the United States, although soda cans are measured using the imperial unit of ounces, soda bottles are measured using the metric system unit of liters.

UNITED STATES

LIBERIA

MYANMAR

Today, only three countries in the world officially use the imperial system of measurement. They are the United States, Liberia, and Myanmar (formerly Burma).

Many unit names come from everyday objects used to measure things hundreds of years ago. The length of a person's foot was used to measure short distances, for example, and a household cup was used to measure liquids. Eventually people set standards. These standards became the British Imperial System. The U.S. Customary System is based on the British Imperial System. Both systems use imperial units.

Consider This

President Gerald Ford signed a law in 1975 stating that the metric system was America's "preferred system" of measure, but most Americans didn't start using the metric system. Why do you think that is?

The Marvelous Metric System

The metric system was invented in France during the French Revolution in the late 1700s. Today, it's used in most countries worldwide. Units of measure in the metric system are based on the meter for length, the gram for mass, and the liter for volume. Each unit is divided into smaller units or combined into larger units by multiples of 10.

The French Revolution, from about 1789 until 1799, made the creation and wide use of the metric system more likely because it caused a time of great change in Europe.

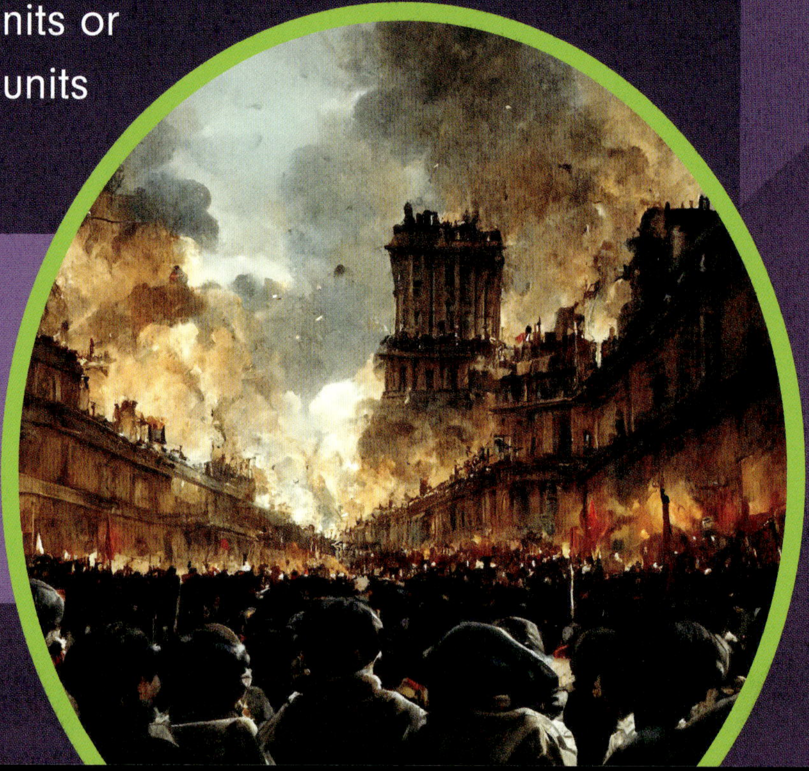

The metric system is well suited to being used by engineers and scientists because it is standardized. It also allows for easier calculations by shifting **decimal** points.

Standard prefixes are added to the basic unit to tell how many multiples of 10 are in a unit. For example, "kilo-" means "thousand," so there are 1,000 meters in 1 kilometer. "Centi-" is a prefix meaning one-one hundredth (0.01) and "milli-" is a prefix meaning one-one thousandth (0.001). So, 1 centimeter is one-one hundredth of a meter and 1 millimeter is one-one thousandth of a meter.

WORD WISE

A decimal is a number written with a dot between the part of the number that is 1 or more and the part of a number that is less than 1. This is similar to using fractions. One-one hundredth is the same as 1/100, which is the same as 0.01.

A Mix of Systems

In 1795, France became the first nation in the world to start using the metric system. Soon after, U.S. Congress considered adopting it. But people were more familiar with the imperial system, so Americans continued to use it. By the mid-1900s, most countries had adopted the metric system and didn't use other measurement systems.

In the United States, the metric system is used in the science, medical, engineering, automobile, electronics, electrical, and manufacturing fields. However, the United States has not adopted the system as its only measurement system. The imperial system can still be seen on road signs (miles, not kilometers, per hour), in cookbooks (cups, not milliliters), and in many construction projects (inches, not centimeters). A 2022 poll found that most Americans prefer to use the imperial system. However, it also found Americans under the age of 45 are more likely to use the metric system.

Speed limit signs in the United States and parts of the United Kingdom are shown in miles per hour, which is part of the imperial system of measurement.

compare and contrast

In what ways are the metric and imperial systems similar and different? What do you think are some benefits and drawbacks of using both systems in the United States?

Most countries in the world use the metric kilometers per hour to measure speed.

Determining Distance

If you're trying to determine the space between two points, you're measuring distance. A ruler can be used to measure short distances. On a metric ruler, a millimeter is the smallest unit. There are 10 millimeters in a centimeter. On an imperial system ruler, inches are commonly divided into sixteenths of an inch.

Most rulers and tape measures in the United States show distance measurements in the imperial and metric units.

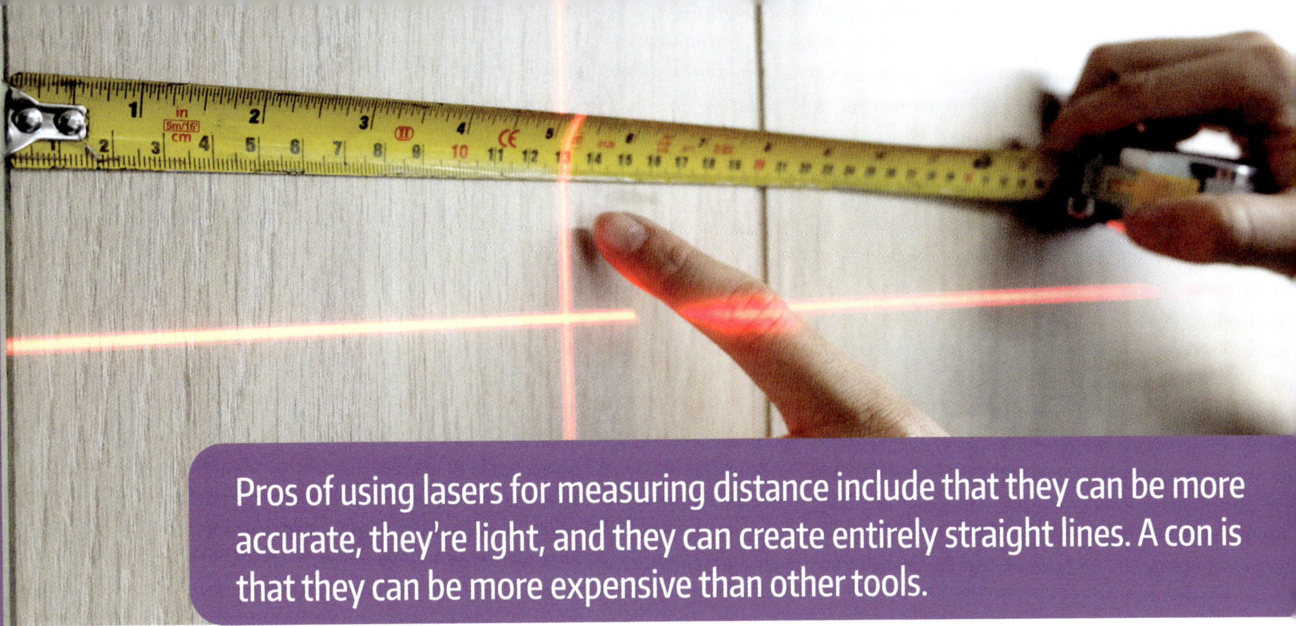

Pros of using lasers for measuring distance include that they can be more accurate, they're light, and they can create entirely straight lines. A con is that they can be more expensive than other tools.

For longer distances, an odometer can be used. In a car or bicycle, an odometer measures distance by counting the number of revolutions (turns) of a wheel. With each revolution, the wheel travels a certain distance forward.

Very long distances can be measured with lasers. A light beam is **emitted** from a laser and reflected back to the source. The travel time of the beam is then measured. When using laser distance meters, this is called the "time of flight" principle. The measurement can be called a "time of flight" or "pulse" measurement.

WORD WISE

To emit is to send something, often light or energy, out from a source.

Amazing Area and Perimeter

When measuring distance, you often only need to worry about a line. However, what if you wanted to measure the distance around a flat rectangular area that has both length and width? This measurement would be the rectangular area's perimeter. For example, to build a fence around a garden, you would measure each side of the garden and then add the measurements up. The length of the fence would equal the sum of the length of the garden's four sides (its perimeter).

Area is the amount of space within a region. It is measured in square units. Imagine laying 1-foot square tiles on a patio so they didn't overlap and there was no empty space between them. The number of tiles used would be the number of square feet, or the area, the patio covers. You can also find the area of square and rectangular objects by multiplying the measurement of the object's length by the measurement of its width.

If you wanted to know how much new flooring to buy for a kitchen, you would need to figure out the kitchen's area, or the space it occupies. This measurement is usually given in squared units.

compare and contrast

What are some similarities and differences between finding an object's area and finding its perimeter? In what situations might knowing both measurements be helpful?

If you walk around the edge of a playground and measure the distance that you walked, you have the playground's perimeter measurement!

Valuable Volume

You've learned how to measure an object's length, area, and perimeter. But what if you wanted to figure out how much it could hold? Then you'd want to find its volume. Volume is the amount of space within an object that has three dimensions. To determine the volume of a box, for example, you would measure its height, width, and length, and multiply these three things. For instance, the volume of a box with 10-foot sides would be 10 × 10 × 10, or 1,000 cubic feet.

Knowing volume is important for most recipes. It allows you to know how much of each ingredient to put into your recipe, from how many cups to how many quarts!

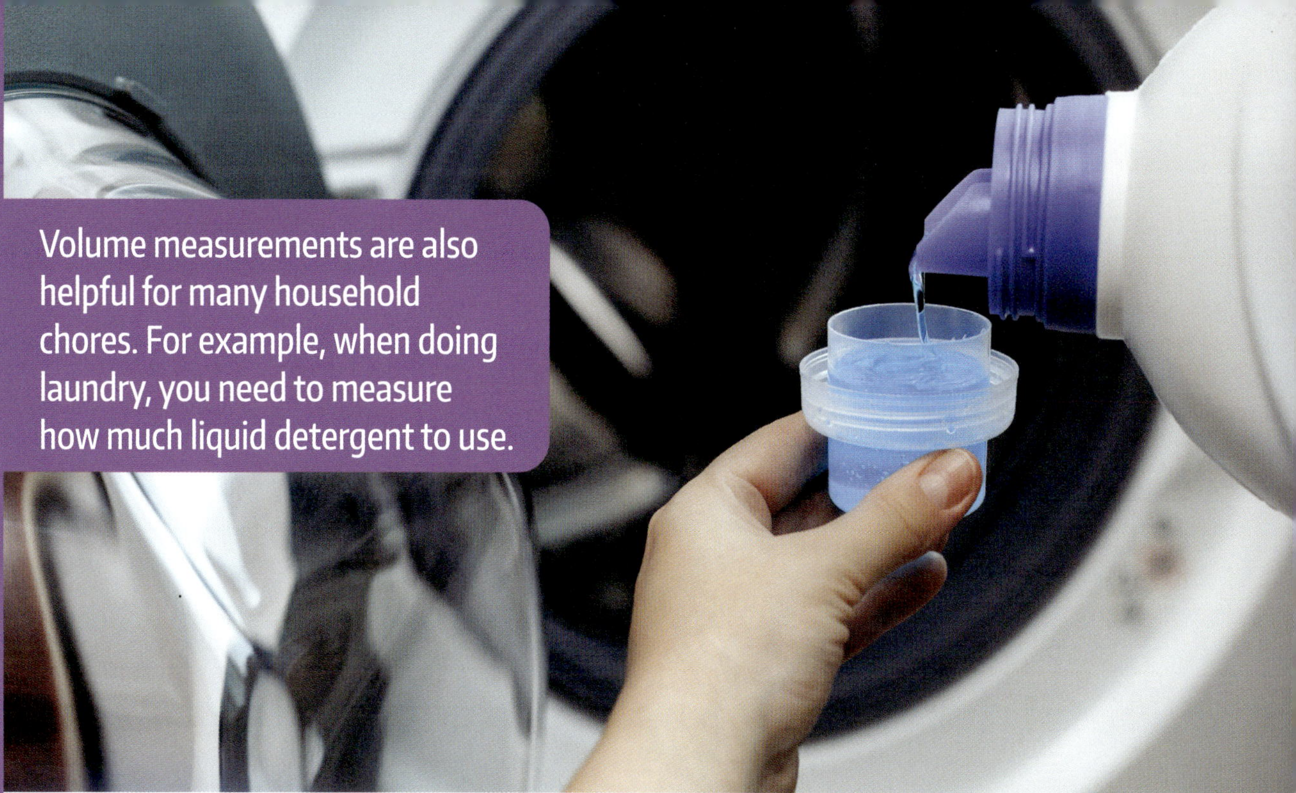

Volume measurements are also helpful for many household chores. For example, when doing laundry, you need to measure how much liquid detergent to use.

To visualize what a cubic unit is, imagine filling the box with smaller boxes, each with 1-foot sides. The number of cubic feet would be the number of smaller boxes that would fit inside the larger box.

Consider This

Volume measurements can also help with conserving, or saving, water. For example, volume allows you to measure how much water you use if you take a 10-minute shower, and how much water you could save by making your shower shorter. Can you think of other ways to use volume to save water?

Measurements for Circular Objects

Circular objects, such as a merry-go-round, have different parts that can be measured. A round object's perimeter, or the distance around its outside edge, is its circumference. The distance across the center of a circle is its diameter. A circle's radius is half of its diameter. It's also the distance from the center of a circle to anywhere on the circle's border. A circle's area is all the space inside it.

What's believed to be King Arthur's historic round table is hanging on the wall of a great hall in Winchester, England. The table has a diameter of 18 feet (5.5 m). What is its radius? What is its circumference? (Hint: Use **formulas** on p. 21.)

The volume ("v") of a **cylinder**, such as a can of your favorite soup, is pi multiplied by the radius squared times the cylinder's height ($v = \pi r^2 h$).

There are formulas for finding each part of a circle's measurement. A special number called pi, which is about 3.1416 and symbolized by π, is used to find a circle's area and circumference. To find circumference ("c"), you multiply pi by the circle's diameter ("d") using the formula $c = \pi d$. Or you can multiply the circle's radius ("r") by two and then by pi using the formula $c = 2\pi r$. The area of a circle is pi multiplied by the radius squared, which is the radius times itself (r x r). This is the formula $a = \pi r^2$.

WORD WISE

In mathematics, a formula is a general rule or fact that is written using symbols and letters. A cylinder is a shape with straight sides and two round ends.

Ways to Measure Weight

There's a force you can't see that pulls all objects towards each other—it's called gravity. The gravitational pull of Earth on the mass of objects is what gives them weight. But most people think of weight as the heaviness of something. In the metric system, weight is measured in grams. One thousand milligrams equal 1 gram. One thousand grams equal 1 kilogram. One thousand kilograms equal 1 metric ton. In the imperial system, the unit for weight is a pound. There are 16 ounces in 1 pound and 2,000 pounds in 1 ton (which is different from a metric ton).

Food scales that measure the weight of certain foods can be helpful for people trying to eat healthier.

Imagine if your home was suddenly moved into outer space! While everything in the house would seem weightless, it would still have a measurable weight. Without any ground below, everything would be in a state of freefall!

Weight is measured using a scale. Since ancient times, balance scales have been used. When using a balance scale, a known weight and the item being weighed are placed on either side of a beam balanced on a fulcrum. When the sides are balanced, it means that the item being weighed equals the known weight. Digital scales are more commonly used today than fulcrum scales, but you may still use a fulcrum scale in a science class!

Consider This

While digital scales are quick and useful, they don't allow you to see how scales work. See if you can make your own balance scale using a few commonplace items, including a ruler to use as a level and a pencil or pen to use as a fulcrum.

Analyzing Angles

What do clothing hangers and scissors have in common? They both have angles you can measure! An angle is the space between two lines that meet at one point. Angles are measured in degrees.

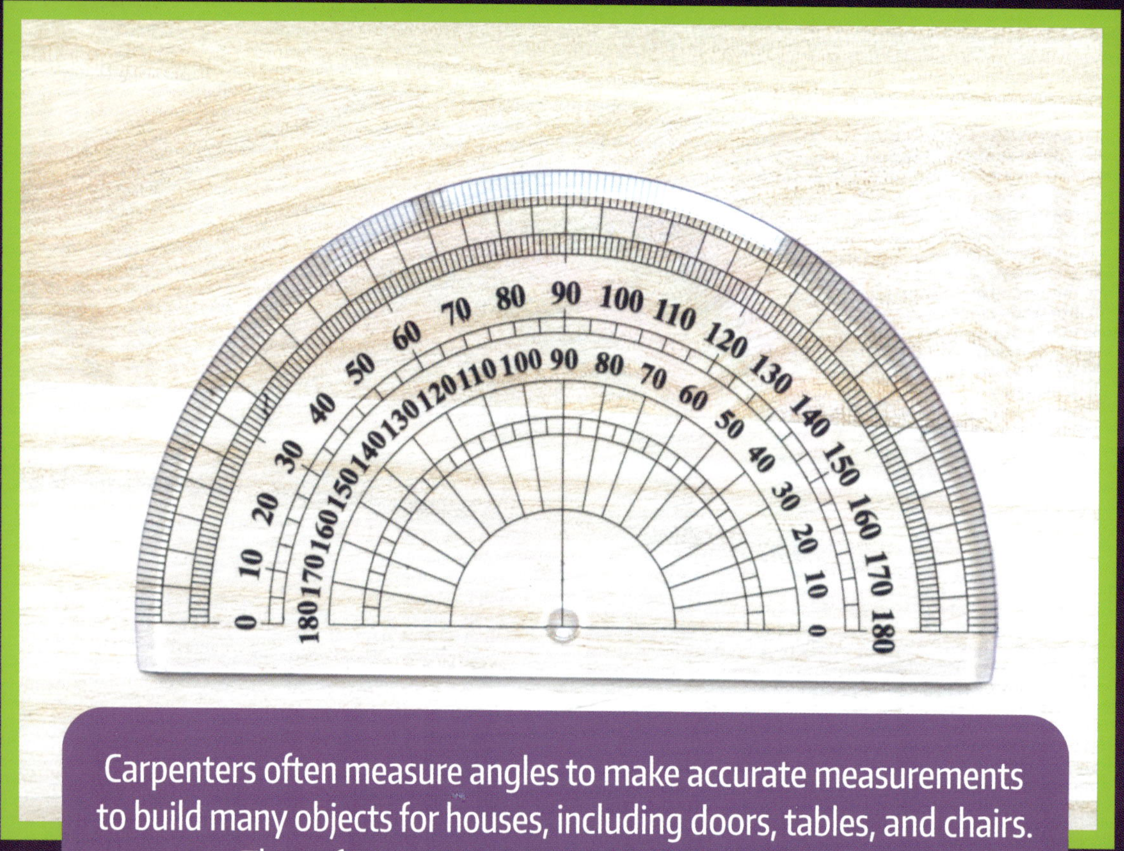

Carpenters often measure angles to make accurate measurements to build many objects for houses, including doors, tables, and chairs. They often use a protractor tool, shown above.

When a laptop screen is standing straight up from its keyboard, it forms a right angle. Opening the laptop wider can form an obtuse angle, and closing it slightly can form an acute angle.

A protractor is the tool used to measure angles. It is often a half circle that has markings along its edge for degrees, which are units for measuring angles. The degrees range from 0 to 180. Laying the protractor's base flat against one line of an angle, the other line of the angle points to its number of degrees on the protractor. A 180-degree angle is known as a straight angle. It looks like a straight line. A 90-degree angle is known as a right angle. Examples of right angles include the corner of a book, room, or window. The outline of all these right angle examples includes one line pointing straight up, and another line lying flat.

Consider This

An angle that measures less than 90 degrees is called an acute angle. An obtuse angle is any angle that measures greater than 90 degrees. Can you think of examples of right, acute, and obtuse angles at home or at school?

How Hot or Cold?

Before heading outside, to know what to wear, you want to know the temperature, or how hot or cold the air is. The metric unit of temperature is a degree Celsius, or °C. The **freezing point** of water is 0°C. The **boiling point** is 100°C. In the United States, the Fahrenheit degree is a common unit. The freezing point of water is 32°F, and 212°F is the boiling point.

The coldest temperature ever recorded outside was 184 K (–128.5°F or –89.2 °C) at Vostok, Antarctica. Scientists have never measured a temperature as cold as 0 kelvin.

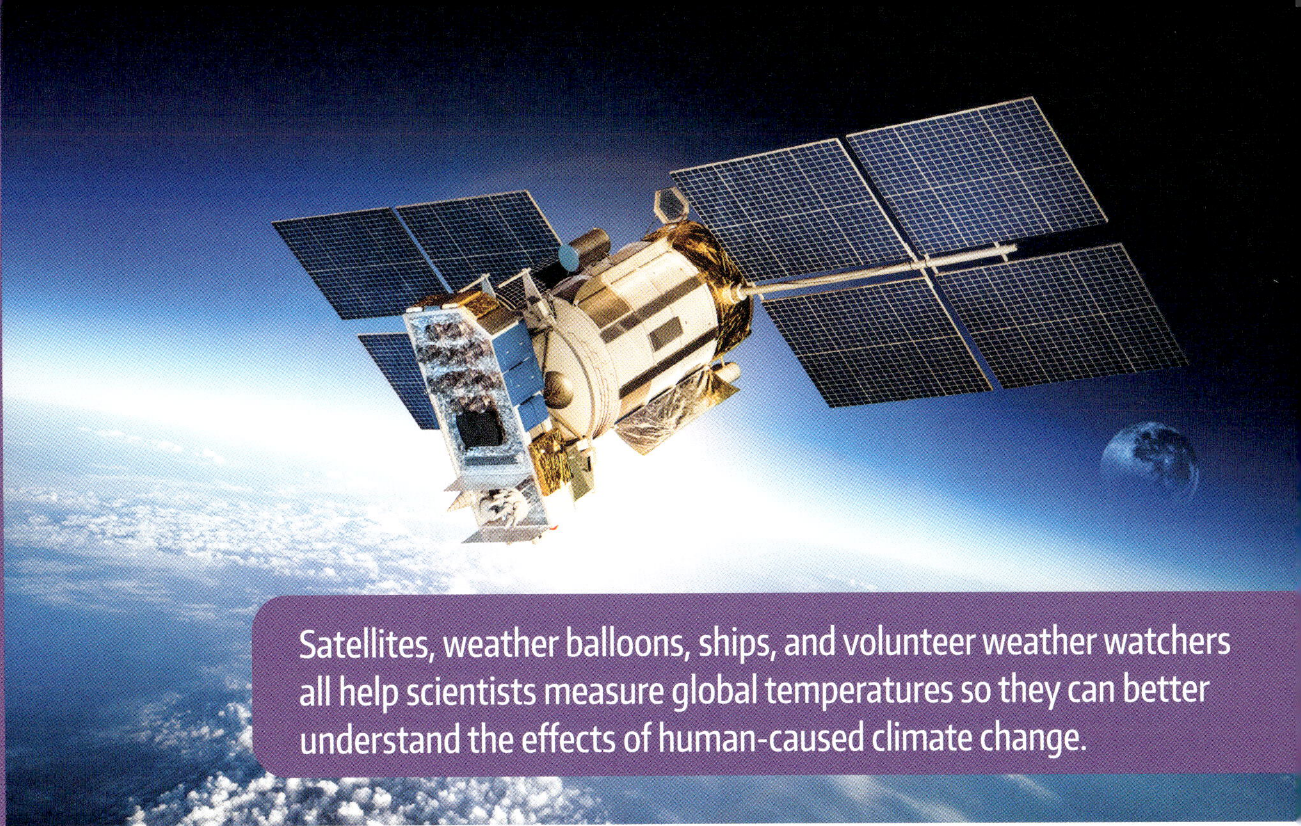

Satellites, weather balloons, ships, and volunteer weather watchers all help scientists measure global temperatures so they can better understand the effects of human-caused climate change.

Another unit, the kelvin (K), does not usually include the term "degree." A single kelvin equals the measurement of a single degree C. This kelvin scale is often used in science because zero kelvin, or 0 K, is called "absolute zero." That means it is the coldest temperature that could ever be possible—much colder than 0°F or 0°C!

WORD WISE

Freezing point is the temperature at which water freezes, or becomes ice. Boiling point is the temperature at which water boils, or becomes so hot that bubbles form and rise to its surface.

Time to Tell Time!

The progression, or order, of events that includes the past, present, and the future is called time. The same units of time are used in the imperial system and metric system. Unlike most metric units, units of time are not based on 10. There are 60 seconds in a minute, 60 minutes in an hour, 24 hours in a day, and 365 ¼ days in a year. Seconds, however, are divided into tenths, hundredths, thousandths, etc.

GPS is short for Global Positioning System. It's a system that uses satellite signals to locate places on Earth.

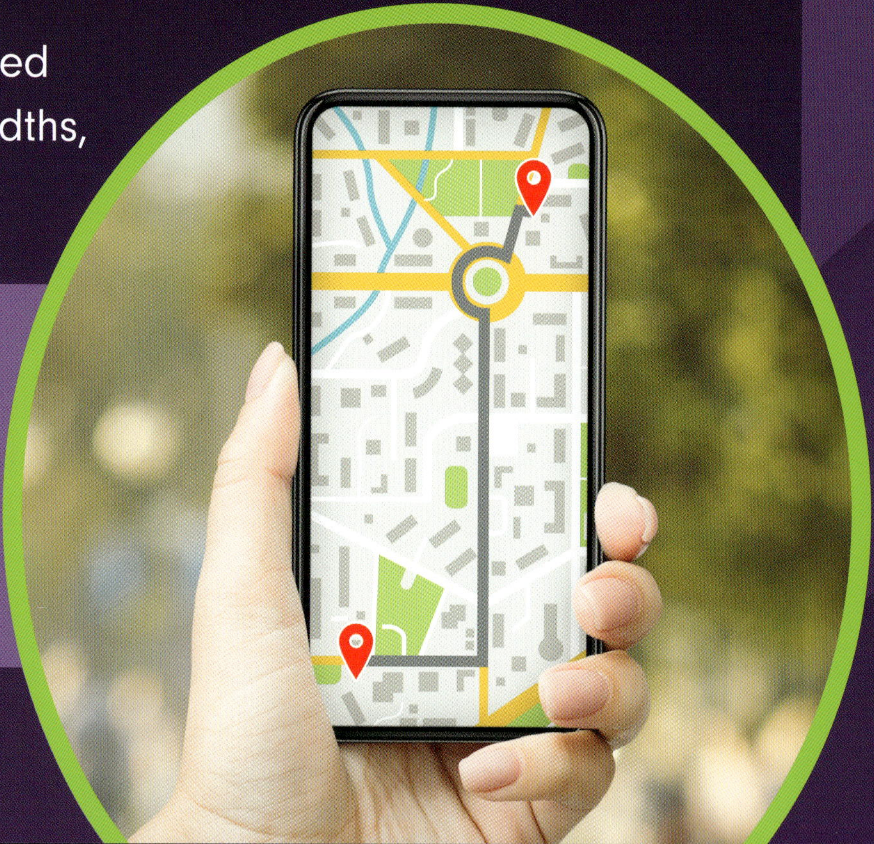

Computer networks synchronize, or line up, their time using atomic clocks. A computer network is two or more computers that are connected to share data electronically.

Time is measured by clocks. Atomic clocks are among the most accurate. Atomic clocks measure time using the **frequency** of the vibrations of certain atoms—usually cesium or rubidium. Atomic clocks have many modern uses, including in GPS systems and computer networks.

WORD WISE

Frequency is the number of times that something happens during a certain time period.

Glossary

climate change Long-term change in Earth's climate, caused mainly by human activities such as burning oil and natural gas.

data Facts and figures. Also, information created and stored by a computer.

detergent A liquid or powder that is used for cleaning clothes, dishes, and other items.

dimension An extension in one direction. We can measure three dimensions: length, width, and height.

engineer Someone who plans and builds machines.

fulcrum The support on which a lever rests.

future A period of time that will happen after the present.

laser A device that creates a narrow and strong beam of light. Lasers are made when electrons in certain materials such as glass, crystal, or gas absorb energy from a light or an electric current.

pi The ratio of the circumference of a circle to its diameter. Its value is approximately 3.1416.

prefix A letter or set of letters that can be added to the beginning of a word to change its meaning.

satellite A machine that is sent out into space and orbits Earth, the moon, the sun, or another planet.

symbol A picture, shape, or object that stands for something else.

vacuum An empty space without any matter in it.

vibrate To move rapidly back and forth.

volunteer A person who offers a service without being paid.

For More Information

Books

Bender, Douglas. *Measuring*. New York, NY: Crabtree Publishing Company, 2022.

Schuh, Mari C. *Measuring the Weather*. Vero Beach, FL: Rourke Educational Media, 2023.

Tyler, Madeline. *Monster Measuring*. Minneapolis, MN: Lerner Publications, 2020.

Websites

Glossary and Terms: Measurement
www.ducksters.com/kidsmath/measurement_glossary.php
Find useful terms related to measuring here.

Measurement Games
pbskids.org/games/measurement/
Test your knowledge of measuring with some fun games!

Why Do Measurements Matter?
www.wonderopolis.org/wonder/why-do-measurements-matter
Learn more about the importance of measurements in your everyday life!

Publisher's note to educators and parents: Our editors have carefully reviewed these websites to ensure that they are suitable for students. Many websites change frequently, however, and we cannot guarantee that a site's future contents will continue to meet our high standards of quality and educational value. Be advised that students should be closely supervised whenever they access the internet.

Index